Abandoned Railway Stations

Stations

A Journey Through Desolate Destinations

Table of Contents

Chapter 1. Introduction

Welcome to a fascinating journey through time and space, as we explore the nostalgic allure of Abandoned Railway Stations: A Journey Through Desolate Destinations. This captivating Special Report delves into the forgotten relics of a bygone era, revealing stories told through crumbling platforms and weather-beaten tracks. You're invited to uncover the beauty in desolation, trace the echoes of history, and unearth the hidden mysteries that lie within these abandoned edifices. Imbued with a quiet beauty and poignant echoes of bygone bustle, these stations bear silent testament to the passage of time. Venture with us on this remarkable expedition, a must-read for history buffs, architecture enthusiasts, and adventurers alike – a journey you truly won't want to miss!

Chapter 2. Dusk of the Iron Horse: An Introduction to Abandoned Stations

In the middle of the 19th century, the Iron Horse, a term coined for the steam locomotive, ruled the transportation industry. Majestic, powerful, and ceaseless, these metallic beasts transported people and goods, forging connections between otherwise unreachable landscapes. Today, their legacy lives on through the skeletal remains of stations they once served - tattered signals, decayed tracks, and crumbling platforms painting a picture of an era now confined to the history books.

Iron was once the heart of a thriving society, with the railway stations standing tall as symbols of progress and innovation. But all that came with an expiration date. As technology advanced, the rise of automobiles and air transportation ensured the gradual decline of these iron horses, leading to the desolation of the railway stations they tirelessly served.

2.1. The Golden Age of Railways

The advent of the Iron Horse brought with it an unparalleled era of prosperity. Railways transformed industries by offering efficient transport solutions, enabling the flow of goods and services. The stations quickly became hubs of activity, bustling with life. The whistle of departure and arrival, the chatter of excited travelers, the cough of the engines, became part of the ever-present orchestra in these spaces.

But the railways did more than simply transport. They were centers of cultural exchange. As trains pulled into stations, they carried with them stories from across the globe, cultural and social threads that wove a diverse tapestry. The stations themselves mirrored this diversity, with their unique architectural styles - from the grandeur of Ruskinian Gothic to the simplicity of the vernacular.

2.2. Falling into Shadow

However, mid into the 20th century, this music started to fade. The introduction of more modern and efficient transportation methods led to the gradual decline of these seemingly invincible iron giants. Cars offered the luxury of personal space, while airplanes shortened travel time drastically. Consequently, railway services began shutting down, leading the architectural marvels, which were once teeming with life, into an irreversible path towards abandonment.

As services began to dwindle, so too did the life linked to these stations. The laughter and chatter gave way to a deafening silence. The vibrant platforms were engulfed in an eerie calm, only disturbed by occasional gusts of wind or rain. These stations were turned into time capsules, holding the remnants from an era long passed.

2.3. Beauty Amid the Desolation

Interestingly, it was this forced retirement that uncovered a new facet of beauty within these structures. Amid the decaying bricks and rusting iron, there lingered a haunting beauty. The battered waiting rooms, deserted platforms, and derelict tracks carry a distinct patina that speaks volumes of their storied past.

Every chip in the paint, every crack in the stone, and every rust stain on the tracks is a testament to the passage of time, like battle scars war-torn soldiers wear. Exploring these sites is like sifting through pages of history, where each fragment plays a significant part in narrating the grand saga of the Iron Horse era.

2.4. The Iron Horse Legacy

While these stations may stand abandoned and forlorn, they remain as silent narrators of a bygone age. In their solitude, they hold the key to an important chapter in human history, where ingenuity was bound only by the limits of technology. They are living museums, offering invaluable lessons in architecture, engineering, and societal evolution.

The legacy of the Iron Horse era is kept alive by enthusiasts and historians who tirelessly work towards preservation and documentation. They venture into these forsaken territories, unearthing hidden tales of life on the tracks. Each visit, each study, reverberates the echo of the steam whistle, reminding the world that the spirit of the Iron Horse still dwells amidst the ruins.

In their desolation lies a sense of melancholy, a poignant reminder of what once was. But there is also a sense of resilience, a stoic testament to the inevitable march of time. The iron horse age might be over, but its memory reverberates through these abandoned stations, standing as silent spectators, solemn reminders of our collective past.

With each exploration, we invite you to delve into this engaging world of steel and stone. Unravel the history, architecture, and stories trapped within these abandoned edifices, and in doing so, celebrate the enduring legacy of the Iron Horse. The tale is written in peeling paint, in crumbling bricks, in rusted tracks – it's a story you definitely won't want to miss.

Chapter 3. Restless Platforms: Stories From the Ghost Trains

From the steamy breath of the Iron Horse to the whirling winds of desolate platforms, the railway stations that once teemed with life have now been condemned to a profound silence. Echoes of laughter and urgent goodbyes, businessmen rushing to catch the 8.15 express, summer picnic outings, first kisses initiated amid the helter-skelter of the platform – all these poignant memories have faded into obscurity, and all that remains is a spectral shell of its former vibrancy. Yet, within these empty stations, ghost trains seem to have their own tales to narrate for anyone patient enough to listen.

3.1. Ghost Trains: A Shadowy Transit

In the realm of the spectral, the ghost train phenomena stands apart. While dismissive skeptics attribute these apparitions to wild imagination or a propensity for the macabre, believers offer tales enough to raise goosebumps, narrations concerning ethereal engines and carriages passing silently through closed stations. Ghostly faces peering mournfully out from behind misty windows, or the quiet hiss of steam giving way to an unnatural silence.

For instance, the antiquated shell of the 'Beckenham Junction', in the heart of London, is steeped in such apparitional narratives. Long since discontinued, midnight walkers swear they still hear the cawing of the stationmaster's whistle, followed by the spectral chug of an approaching train. Yet, when they turn, they see only the remnants of the old platform, elegant in its worn grandeur, but empty of any tangible presence.

3.2. Silent Witnesses of History

Abandoned railway stations can sometimes serve as a silent witness to significant historical upheavals. This is profoundly felt in the skeletal remains of 'Warschauer Straße', in Berlin, Germany. During the days of division, the station was forced into a premature retirement, being located precariously close to the Berlin Wall. After the Wall collapsed, 'Warschauer Straße' was gradually reclaimed by nature, its tracks surrendering to weed, and its platforms serving as roosting spots for birds.

Now, if you were to venture to this desolate place on a foggy morning, you'd feel a shiver of something unexplainable, a vague remnant of herded men and women, of shouts that were strangled before they found voice, a heavy silence pregnant with whispers of suppressed stories.

3.3. Abandoned to Art

In an intriguing turn of events, some abandoned train stations have traded their practical function for a creative one, becoming the canvas for graffiti artists. New York's 'Freedom Tunnel', previously a neglected stretch of rail beneath Riverside Park, is a celebrated example of such transformation.

The tunnel, named after the graffiti artist Chris "Freedom" Pape, who found inspiration in the forgotten remnants of the tunnel, is now home to some stunning murals that have transformed the desolate route into an underground art gallery. The artwork serves as poignant commentary, capturing the essence of urban decay, social ignorance, and the often overlooked beauty of entropy.

3.4. Conclusion: Fading Tracks to Yesteryears

The stories of the restless platforms and ghost trains serve as a melancholic testament to the beauty of hope and the inescapable encroachment of time. These abandoned railway stations, once hubs of hustle and bustle, now stand mute, their silence disrupted only by the ethereal sounds of phantom trains and spectral passengers. But within their crumbling façades is a recorded history of human endeavor and resilience, an echo of the indomitable human spirit.

Every platform scarred by the attrition of time, every weather-beaten track, every fading mural painted on a lonely tunnel wall captures a fascinating story, a narrative that lies hidden in the heart of desolation. Thus, they continue to present an inexhaustible allure for historians, adventurers, and lost souls alike, inviting them to tap into the silent echoes and tune into the whispering tales of these skeletally beautiful fragments of history.

Chapter 4. Art of Ruin: Architectural Splendor of Derelict Stations

The tranquility of abandonment holds a discreet charm. It narrates tales of past rambunctiousness and the gradual fade into silence. The architectural beauty of derelict stations amidst this tranquility is a sight to behold. Time-worn designs offer a discernible aesthetic, treading the fine line between decay and grandeur.

4.1. Architecture as Living History

Railway stations have always been more than just transport depots. They are living fragments of history and culture, mirroring the evolution of civilization and technology. Architecturally, they reveal a fascinating mix of styles, influenced by the era they were built in and the vision of their architects.

The glorious facades of these forgotten edifices are reminiscent of varying styles from the vernacular to Art Nouveau, Neoclassical to Brutalist, and much more. Their designs unveil an intricate amalgamation of practicality and aesthetics, honed over decades of architectural evolution.

However, the abandonment of these places imbues them with an unexpected beauty, as architecture and nature merge, lending the structures an organic feel. Moss-covered platforms, ivy-laced columns, and weather-worn walls paint an enticing canvas of the inevitable dance between human creation and natural decay.

4.2. Grandeur amid the Ruins

Exploring further, one stumbles upon erstwhile grand terminals and stations that once teemed with life, now surrendered to desolation. Their crumbling architecture, adorned with intricately carved arches, tall columns, and ornate moldings, speak volumes of their prior grandeur.

The Michigan Central Station, once the bustling heart of Detroit, is a prime example. Despite years of neglect, it retains an imposing silhouette — a melancholic ode to the city's once-thriving industrial past.

Across the pond in Europe, the Canfranc International Railway Station in Spain stands as an impressive example of Spanish Renaissance Revival architecture. Despite its disuse, it continues to captivate onlookers with its stonework and gargantuan size, a haunting testament to the opulence of the past.

4.3. Stations as a Reflection of Societal Progress

Regardless of their condition, these stations offer a window into the social and technological progress of their operating times. They were not merely buildings, but tangible symbols of the Industrial Revolution, attesting to the rise of technological prowess and human innovation.

The structures brilliantly incorporate technological innovations prevalent during their construction. Iron and steel frameworks, coherent use of glass, and advancements in structural loading are visibly evident in their architectural treatment.

The void of activity further amplifies these elements, as the noiseless environment steers focus towards these details. The bare, stripped-back character of the space uncovers these stories, bearing testimony to the roots of our current society.

4.4. A Testament to Resilience

Despite the ruinous state, these stations emanate an incredible sense of resilience. Battle-scarred by years of exposure to elements, and yet remaining in place, they radiate a silent defiance against the passage of time.

The broad-gauge track formations, the extensive platforms, the rugged iron pillars take on new life as art forms, lending the stations a unique identity that awakens a sense of admiration, and often reflection, in the observer. Amid their decay, they expose the beauty of endurance and inspire conversations about the transience of human existence.

4.5. Rebirth through Dereliction

Remarkably, the desolation of these stations has given birth to a new form of appreciation: 'ruin aesthetics' or 'beauty in decay.' This perspective celebrates the decomposing features of the structures as abstract art, garnering significant attention from artists, photographers, and architecture lovers worldwide.

Their ruins are turning into open-air galleries, displaying a splendid cocktail of architectural remnants, graffiti art, and nature's reclaim. The silent platforms, lonesome benches, and fading signboards parallel contemporary art — evolving continuously and unpretentiously.

As we tread through this journey, the potent allure of the desolate stations calls for a deeper look at our shared past. Each station is a reservoir of untold stories, architectural innovations, and historical periods, narrating a beautiful dance of permanence and impermanence, a snapshot of history frozen yet evolving in time. Despite their abandonment, the stations flourish as 'Art of Ruin,' boasting a subtle network of aesthetics, sentiments, and complexity.

Chapter 5. Rust and Revival: Conservation Efforts for Historical Monuments

Efforts to salvage vestiges of the past form a significant, albeit often unheralded, chapter of human endeavor. In this particular pursuit of preserving the past, abandoned railway stations take center stage, providing a slice of history through their peeling paintwork and decaying structures.

5.1. The Urgency of Conservation

The importance of conserving these abandoned stations is multi-fold. To begin with, these stations are tangible remnants of our shared heritage. Each brick, each faded signboard, and the rusting iron tracks carry in themselves fragments of our collective past. They are reminders of the beginning of numerous journeys and the endpoint of many more. Their preservation is thus not just an act of honoring our past but also a way to keep a piece of our history alive for generations to come.

The architectural significance of these stations is another cogent reason for their preservation. Dating back to the era of the Industrial Revolution, many of these stations exhibit unique architectural styles indicative of their period of construction. From neoclassical to art deco, the variety of styles reflects the aesthetic evolution and building practices of different epochs, making these stations invaluable relics of architectural history.

Furthermore, by repurposing these deserted structures, we can put them to innovative uses that can benefit communities. Many defunct stations have been resurrected as museums, restaurants, or community centers, invigorating the local economy while preserving a sense of historical continuity.

5.2. Pioneering Projects: Rehabilitating Railway Relics

The task is undoubtedly gargantuan, and various professional and grassroots organizations worldwide have taken it upon themselves to breathe life back into these crumbling structures.

In England, for instance, the Railway Heritage Trust undertakes restoration works on numerous disused stations throughout the nation. Its most notable work perhaps being the restoration of the Alton Railway Station in Hampshire, which was deftly transformed into a bustling public hub, housing a theatre, a restaurant, and a community center, each intriguingly ensconced within the charming Victorian structure.

Crossing over to North America, the California State Railroad Museum Foundation has been instrumental in preserving California's rich railroad history. It has adopted several abandoned stations, converting them into museums that now serve as vibrant educational resources, offering us valuable insights into the heydays of the American railroad industry.

Chapter 6. Lessons from Successful Revivals

The successful revival of once-forgotten stations hinges on a combination of factors. Foremost among them is understanding the architectural and historical relevance of the structure. Integral to this process is detailed archival research that includes studying historical photographs, architectural plans, and railway schedules, among others.

Equally crucial is an inclusive approach that engages the local community. Encouraging public involvement generates greater awareness, interest, and often, crucial financial support. The successful rebirth of the Tarkio, Missouri station as a vibrant civic center stands as a testament to the power of community-led conservation endeavors.

6.1. Challenges and Strategies

Despite the successes, challenges abound. The lack of funding and public apathy are the biggest hurdles. For this, tactical alliances with government bodies, private enterprises, and influential individuals can secure both financial and administrative aid. Moreover, a well-implemented public awareness campaign can go a long way in mobilizing widespread support.

6.2. The Road Ahead

The path to conservation is fraught with difficulties, but the gains unequivocally outweigh the challenges. As we chart our course ahead, we need to continually adapt and evolve our strategies, building resilient partnerships and harnessing emerging technology to ensure the survival of these silent sentinels of our past.

In sum, the conservation of abandoned railway stations presents us an extraordinary opportunity to save links to our shared heritage from oblivion. As we breathe life back into these structures, we keep history alive for generations to come. Through strategic efforts, pioneering projects, and concerted action, we can ensure the rust of disregard gives way to the sheen of revival - allowing these abandoned railway stations to stand proud, against the relentless tides of time.

Chapter 7. Stations of the Lost: A Closer Look Into Closed Track Lines

Time halted its merciless march for these hauntingly beautiful relics of a quintessential bygone era. Closed track lines bear witness to the transformative shifts of our world, from both social and technological dimensions. Standing as silent spectators to the rhythms of human life, they epitomize resilience emanating from abandonment, evolving expectations, and the relentless tide of time.

7.1. The Last Departure

The tale of abandonment often commences with the final departure. Picture the quiet station, awaiting its last train, filled with unsuspecting passengers embarking on a journey, oblivious of an era nearing its end. The platform vibrates one last time as the locomotive rumbles into view. Luggage is hurriedly loaded, goodbyes hastily whispered, and lives duly intertwined in the bustling narrative that is human life. And as the last carriage disappears into the distance, an era quietly folds up behind it, leaving an architectural skeleton in its wake – yearning for use, yet destined for desolation.

7.2. Frozen Moments in Architecture

Integral to these closed track lines is the architecture. Every ticket booth and rotting wooden bench intertwines a unique tale of craftsmanship with a melancholic narrative of obsolescence. Fondly interacting with brick, wood, and iron, these structures paint a poignant architectural testimony, rich in character, design, and detail. The blending forms, materials, and designs offer a potent testament to human endeavor and artistic vision, standing tall as living museums of a time when steam, steel, and sheer will propelled mankind cross vast distances.

7.3. The Overgrown Enigma

Surrounding the deserted stations and closed track lines await Mother Nature, ready to reclaim her territory. Initially, an eerie silence engulfs these abandoned places before nature orchestrates her peaceful symphony. Birds return to their forgotten nests in boarded-up stations, and vines snake through the intricate designs of ironwork, softening harsh lines drawn by mankind. Interestingly, the interaction of nature and human architecture recites a simple yet profound tale – the persistence of life and the inevitability of reclamation.

7.4. Tracks to Nowhere

The spectacle of closed track lines, weathered rails being swallowed by the encroaching wilderness, indeed elicits an unsettling sense of intrigue. The eerie serpentine pathways delving into oblivion showcase an uncanny beauty that bewitches the beholder. These twisted, rusty lines, once the arteries of the industrial world, now serve as cryptic memento of human history, scarred by seasons, solitude, and entwined with countless forgotten journeys.

7.5. Tales of Forgotten Glory

Every closed track line presents a silent soliloquy of its past being. While some whisper tales of bustling industry and thriving trade, others recount narratives of crowded platforms and the rhythmic cacophony of steam engines. Echoes of arrival and departure announcements, the cacophony of livelihoods made and lost, and snippets of everyday human melodrama still hang in the cool, forgotten air of these silent stations. The sense of nostalgia is palpable, evoking a wistful yearning for a time lost to the annals of the past.

7.6. Mysteries and Urban Legends

Abandoned railway stations often serve as a breeding ground for unsolved mysteries and urban legends. Steeped deeply into the folds of local lore, they stand as silent witnesses to decades of stories and tales. Whether it be a spectral passenger awaiting a train that will never arrive or an eerie melody floating in the still midnight air, these fascinating tales of folklore underscore the enigmatic charm and ageless intrigue that surrounds these desolate locations.

7.7. Rehabilitation and Transformation

Interestingly, while some of these ghost stations thrive in desolation, others have witnessed a rebirth. From housing trendy cafés to transforming into bustling movie sets, these places embody the theory of creative destruction. They serve as an enduring testament to mankind's ability to adapt, reimagine, and breathe new life into structures that previously bore witness to a bygone era.

As we draw the curtain on the time-worn narratives of these closed track lines, we must take a moment to ponder. These forgotten railways and abandoned stations serve not just as stoic reminders of a glorious past; they hold a mirror to our present, urging us to navigate wisely through the incessant passage of time. As we journey forth through life, let's revere these remnants of our history that surrender themselves to silence, coaxing us to unearth their stories, mysteries, and beauty hidden beneath layers of dust and time.

Chapter 8. Railway Ghost Towns: Exploration of Abandoned Railway Communities

Railways were once the lifeblood of communities, reaching even the most secluded corners. Akin to the arteries pulse through our bodies, trains conveyed a network of vibrant life, beating heartily with the rhythm of industry and commerce. Yet, with the passage of time, what once buzzed with activity now lies dormant, as if stuck in a profound and eternal silence. Dwelling in this eerie tranquility are the ghost town stations of the railroads, remnants of a vanished age.

8.1. A Brief History

The railway was first introduced in the 19th century, at the height of the Industrial Revolution. For the towns that sprouted around these stations, the railroads became the driving force of the local economy and society. However, the advent of automobiles and motorways in the 20th century drastically altered the transportation landscape. Passenger trains gave way to cars and trucks, leaving countless railway communities out in the cold. Gracefully dismissed from our world of rush and bustle, they receded into obscurity, becoming fragments of our collective memories.

8.2. The Shadows of Train Stations

Train stations were once the fulcrum around which local life revolved. Today, many stand desolate, whispering tales of days gone by. Boarded-up windows stare blankly at rusted tracks; dilapidated platforms patiently wait for passengers that never arrive. Here, the clock doesn't tick - it simply continues to echo the last strike, waiting to be wound up once again.

Penn Station, Detroit, is a fine example. Once bustling with travelers, merchants, and goods, it is now akin to a grandiose mausoleum. The handsome architecture still bears the signature of a prosperous past, yet the vacant tracks and deserted corridors offer an eerie testimony to its current state of abandonment.

8.3. The Pulse of Vanished Communities

Transportation fueled not just the movement of goods and people, but also the very essence of societal life. Towns that formed around these stations were throbbing with vitality, with the railway setting the tempo. However, the cessation of rail services would often mean their implosion, pushing once-thriving communities into the realm of ghost towns.

Thurmond, West Virginia, is a poignant example. As a booming coal town at its peak, Thurmond boasted a healthy population, thriving businesses, and, at one point, the largest revenue on the Chesapeake & Ohio Railway. However, as coal fell out of favor, the town dwindled. Today, it stands with around five residents, testament to the caprices of time and industry.

8.4. Tale of Railway Hotels

Railway hotels, once synonymous with comfort and luxury for long-distance travelers, now stand as relics of architectural beauty. The rise of the motorcar era caused their slow decay, and many of these magnificent structures have slept untouched for decades.

The Canfranc International Railway Station, nestled in the Pyrenees between Spain and France, is an incredible example of ornate grandeur left to decay. With 365 windows and over 200 doors, it was once known as the 'Titanic of The Mountains'. Today, its deserted halls and vacant rooms echo an era of luxury and elegance long gone.

8.5. Preserving the Ghosts of the Past

Though abandoned, these towns speak volumes about our history, offering rich insights into the evolution of society, politics, and technology. Preservation efforts are underway, with many deserted stations finding new life as museums, heritage centers, or tourism hotspots. This renaissance, led often by local communities and enthusiasts, keeps the past alive, allowing future generations to witness these monuments to another age.

Thus concludes the chapter: ghost towns straight from an era that time seems to have swallowed up. These disappearing relics are resounding reminders of how the pace of human innovation can leave behind footprints of derelict grandeur. They invite us to contemplate the relentless passage of time, urging us to preserve and celebrate these memories etched in rusty tracks and weather-beaten platforms. While the ghost towns remain deserted, they are far from forgotten, and the stories they tell are far from over.

Chapter 9. Retracing Routes: The Economic Impact of Railway Abandonment

From the inception of railways in the early 19th century, they rapidly became pivotal nodes in the national and international economic networks. The hauling of freight and passengers over vast distances was revolutionized, supporting industrial growth and urbanization on an unprecedented scale. However, along with the inevitable descent of some railroad lines into disuse and abandonment, came substantial socio-economic impacts that transformed the landscapes they once traversed.

9.1. The Rise and Fall of the Railways

The rail networks that sprang up across Europe and the United States in the 19th century were some of the most ambitious engineering projects of the era. These steel arteries acted as powerful catalysts for regional economic development. Towns and cities blossomed around stations and depots, fueled by the industrial might and commerce the railways brought.

By the mid-20th century, however, the heyday of the rail networks had passed. Cars, trucks, and airplanes had become the preferred mode of transport for people and goods. Railways were seen as slow and outdated, and their operations became financially untenable. As a result, some lines were abandoned, leaving the stations deserted and the tracks to decay. The economic consequence of this abandonment was soon felt, with local industries and communities suffering losses.

9.2. Local Economic Impact

When the trains stopped running, the economies of the towns that had grown up around railway stations suffered. Industries that depended on the railroads for transportation of goods found their profits diminishing. They either had to invest heavily in road transport or face closure. As factories began to close their doors, unemployment in these railway towns skyrocketed. It was the start of a downward spiral from which some communities are still trying to recover.

There was a ripple effect that traveled through these towns, affecting local services from shops to schools. The loss of a stable income in families meant lower consumer spending, affecting local businesses. This, in turn, resulted in a loss of tax revenue for local government services including education, healthcare, and maintenance.

9.3. Decline of Infrastructure

The decline and ultimate abandonment of the railways led to another economic blow: a deteriorating infrastructure. Leaving these architectural wonderworks - stations, bridges, viaducts - to suffer the ravages of time was a loss of valuable capital. Moreover, maintaining these structures in a non-degrade state required funds that many local governments simply did not have.

In addition, the abandoned tracks often severed transportation routes. So, even when localities had the potential for growth, these railroads represented an almost impenetrable barrier to progress. They divided towns, reduced accessibility, and hindered future development.

Meanwhile, in the cases where disused railways were converted into trails, while they become valuable recreational spots for biking and walking, the economic impact often varies. These conversions, while positive in terms of environmental and health aspects, do not always contribute to local economic prosperity.

9.4. The Perverse Silver Lining

In a perverse twist of fate, the abandonment of some railway lines has brought about a slow-burning but real regenerative economic effect. New opportunities have emerged from the ghostly remains of these once active community lifelines.

Tourism, driven by the historic and romantic allure of disused railways and the stunning beauty often found along these routes, has brought some communities unexpected prosperity. Abandoned railroads, repurposed into hiking trails, drive significant tourist footfall and rejuvenate local economies.

For some regions, history has come full circle with the railways getting a second lease of life as trendy areas of heritage tourism. Breweries, museums, and entertainment venues have sprung up in refurbished station buildings, signaling a second life for these historic buildings. This has resulted in job creation and has attracted both tourists and new residents, leading to community revitalization in some areas.

9.5. The Role of Policy and Planning

A shift in policy and planning offers the key to maximizing the potential economic benefit of these abandoned sites. It requires a coordinated approach between local communities, policy-makers, and private companies to transform these derelict railroads into engines of economic growth.

Preservation efforts can serve as an economic catalyst, creating construction jobs and attracting tourists. While the challenge is substantial - reviving stagnant economies, maintaining decaying infrastructure, circumventing potential environmental contamination issues, and resolving complex legal problems - the potential benefits are immense.

Abandoned railways are testimony to the economic rise and fall of industries and regions. They represent a past where steel rails were lifelines to prosperity. Today, they stand as silent relics, but with public-private partnerships and innovative transformative policies, they could once again play a part in the economic future, serving as vivid reminders of the resilience and continuous evolution of economies and societies.

Chapter 10. Ghosts of Industrial Revolution: Sociocultural Perspectives

The birth of Industrial Revolution in the 18th century bloomed railways as the beating heart of industrialization. The iron veins spread across the nation, transforming rural landscapes into booming industral towns. Although a cycle of abandonment later swept into these railway stations, a deeper delve reveals much more than dereliction; they whisper tales of sociocultural transitions, technological advancements, and shifts in industrial practice.

10.1. The Age of Steam and Steel

The introduction of steam-powered locomotion radically reshaped the pace and the scope of life during the Industrial Revolution. The railway stations of this era were symbols of human verve and scientific exploration, architectural havens of innovation, and hustling, bustling hives of activity that epitomized the revolutions in transportation and industrial output. Famed engineers and architects endeavored to create not only functioning spaces, but also cultural milestones and architectural marvels.

What we now see as ruins were once the hubs of activity – environments alive with human emotions, animated by hurried footsteps, anxious glances at watch faces and the cacophonous symphony of steam engines. With developments in steel forging paving new avenues of architecture and engineering, smoke-clouded skies of these stations bore witness to societal shifts, when rural peasantry transmuted into the industrial working class.

The swift industrialization, mechanized production and heightened mobility brought on by the advent of railroads also led to a large-scale urban migration. The network of stations often abutted diverse neighborhoods - from the lavish Victorian upper echelons within city precincts to the modest outskirts often populated by railway employees. Over time, these stations developed their unique socio-regional identities, often reflecting the microcosm of local culture, economic status, and lifestyle.

10.2. Transition to Electric and Diesel

By the turn of the 20th century, rails began to hum to the beat of electric and diesel engines. What once was deemed modern became obsolete, and steam engines, the titanic beasts of the industrial anthem, were relegated to history. Railway stations serving these locomotives faced their demise, as they no longer kept pace with the transforming momentum of technology and society.

However, this sweeping change also resulted in a tapestry of fascinating dichotomy. Some stations emanated the old-world charm of steam engines while others mirrored the geometric severity of machine age aesthetics. These architectural fossils portray a unique period of overlap – when old traditions grappled with the new norm, offering a window into the tumultuous transformation society underwent during this period.

10.3. The Ghost Stations and Changing Sociopolitical Climate

Warring times often witness an intensified focus on mobility and communication, surfacing during the World Wars especially. Railway stations were no exception. Many became strategic sites, while the wars' aftermath left others deserted. The changing geopolitical status, coupled with the fall of empires and shifting borders, further led to the abandonment of numerous railways. These 'Ghost Stations', casualties of sociopolitical tides, stand as architectural tombstones of turbulence past.

Why should we care about these deserted edifices of yesteryears? Because they are more than mere architectures; they are time capsules encapsulating individual histories, narratives of communities, technological evolution, and societal progress. The weather-beaten tracks and silent platforms are echoes of eras past, echoing tales of struggle, resilience, change, and hope. Each station carries not just the weight of its brick and mortar, but also the imprints of countless footprints – each a tale of human experience etched in the passage of time.

10.4. Revival and the Cultural Renaissance

The true essence of these stations' historical worth, however, lies not in their abandonment but in their potential for revival. They serve as canvases upon which narratives can be reimagined, histories reinterpreted, and spaces repurposed.

Across the globe, efforts have been made to re-purpose these once-important hubs of societal progress. Abandoned railway stations across Europe, East Asia, and the Americas have undergone transformations — some into art galleries or urban farms, others into community centers or transportation museums. They have evolved into spaces of cultural interpretation and sociocultural convergence, fostering the growth of new communities while retaining echoes of the old.

By reviving these nodes of history, we breathe life into their stories, giving them a platform to be retold and reinterpreted through the lens of modern society. Elements of historical preservation paired with modern innovation can instill community pride, spurring cultural renaissance, and contributing to the broader narrative of development and progress.

In the end, these deserted railway stations are more than relics of iron and stone. They are testaments to resilience in the face of shifting societal and industrial landscapes. Even in their apparent silence and abandonment, they remain a significant sociocultural symbol - providing valuable insights into their past, drawing attention to the present state of urban and rural development, and pointing towards future possibilities of regeneration. Walking through these stations is akin to journeying through time, tracing the patterns of human history, turning pages of a compelling story etched on rust, iron, and stone.

Chapter 11. Nature Reclaims: The Ecological Impact of Railway Abandonment

As mankind ebbs and flows, Nature inevitably rushes in to fill the void. Such is the case with the sprawling networks of obsolete railways, their steel and stone structures gradually reclaimed by the flora and fauna they once disrupted. An unexpected ecological benefit has been observed as these rail corridors are deserted and consequently enveloped and repurposed by the local environment. Let's delve deeper into this wonderful yet complicated process of natural reclamation.

===Rails to Trails: A New Path Forward

In several parts of the world, abandoned railway corridors have given rise to the "Rails to Trails" movement, turning disused tracks into miles-long stretches of hiking, biking, and equestrian trails. These paths zigzag through urban, suburban, and rural areas, uniting communities and offering safe pathways for non-motorized transportation and recreation.

In addition to their value as human thoroughfares, these railway-turned-recreation paths serve an essential ecological purpose as well. They form linear greenways, essentially narrow strips of protected land that put up an avenue for the movement and growth of wildlife and plant species, effectively reestablishing habitats and migration corridors that the original railways might have fragmented.

The vegetation flourishing in these trails represents local flora and sometimes rare or endangered species, providing a rich source of bio-diversity often starkly contrasting with the surrounding urban or agricultural landscapes.

Encroaching Echoes: Mother Nature Strikes Back

In more remote areas, where old rail lines remain untouched, the process of reclamation can be staggeringly rapid and dramatic. Left to the mercies of the elements, railway lines, stations and ancillary structures become strangely eerie tableau; their age and abandonment marked by encroaching greenery and wildlife.

Here, the rusted tracks, weathered wooden ties, and crumbling platforms aren't just picturesque artifacts. They also serve as a unique microcosm of reinhabitation. Signal boxes are claimed by nesting birds, isolated platforms become homes for shrubs, and wildflowers bloom along the forgotten lines, wiping away the metallic residues of industrialization.

The track ballast, comprising crushed stones originally intended for improved drainage, now presents a rocky substrate. This peculiar environment is suitable for many pioneer species which specialize in colonizing disrupted or damaged ecosystems, kick-starting the process of ecological succession.

A Safe Haven: The Railway Ecotones

Older, more well-established railways traversed bustling towns as well as wildernesses, often severing habitats and causing major disruptions to local ecosystems. However, now in their absence, these 'railway ecotones' have become corridors for various forms of life, acting as vital links between fragmented habitats.

The resulting mosaic of different environments along the track – dry, sunny, sheltered, exposed – offers a diversity of microhabitats which can support a wide range of flora and fauna, further enriching the local biodiversity.

Moreover, these defunct rail lines often follow more natural and wildlife-friendly courses than their replacement infrastructure, such as highways and industrial complexes, which are more disruptive and perilous for native species.

11.1. Industrial Skeletons to Wildlife Sanctuaries: Case Studies

Around the world, there have been numerous famous cases of wildlife and plants overtaking abandoned railway lines. Berlin's Südgelände Nature Park is one such example, a former marshalling yard transformed into an urban wildlife refuge, home to more than 100 species of birds and a variety of microhabitats.

In South America, a section of the Antofagasta and Bolivia Railway reaching from Chile to Bolivia has been completely reconquered by nature. Here, endangered vicuñas can be found huddling in deserted station buildings for shade and shelter, while desert-adapted species of succulents and grasses thrust through the arid ground along the rails.

Similarly, the High Line in New York City reveals the power of transformation from rusty rails to a stunning overhead park, demonstrating the potential of these abandoned spaces both for urban rejuvenation and ecological conservation.

In conclusion, the abandonment of railways, while marking the end of one historical and industrial phase, signifies the advent of an intriguing ecological phenomenon. Their derelict lots, platforms and lines, once bustling with human enterprise, depict an abrupt transition, but, in this silence and neglect, an ecological uniformity is emerging. From walking trails to wildlife corridors, the environmental transition is not just impressive, but also an unintentional testament to humanity's ability to reshape the environment, both for better or for worse.

Chapter 12. Lessons Learned: Modern Transport Infrastructure and the Ghosts of the Past

There exists an inherent tension in time's tandem dance with progress and decay. Modern transportation infrastructure, pulsing with energy and activity, is a testament to technological advancement. Yet, among the steel, concrete, and dashing metropolis, stand the forsaken specters of the past, many in the guise of abandoned railway stations, left to narrate their tales through silence and solitude.

12.1. The Dawn of Railways

Railways were born out of an industrial ambition — a vision to create a structure that would connect us more closely than ever before. They represented progress, the triumph of human engineering over geography's relentless constraints. In the 19th century, the steam engine's rhythviric pulse and the track's triumphant unfurling across landscapes epitomized movement, connection, and the unassailable march of human progress. Mankind had birthed a beast of iron and steam that could traverse untold miles, linking cities, bridging nations.

But as technology galloped ahead, many stations were left in its wake. Left derelict, they became stagnant pools in the flowing river of time, still in their desolation, whispering tales of what once was their glory.

12.2. Abandoned Stations: The Ghosts Amongst Us

Visiting an abandoned railway station is a unique experience. The air seems denser, time treads more lightly, the sound of the wind whistles like spectral echoes of bygone trains. Such places are museums of our engineering history, encapsulating in their decaying architectures both the progress made and the progress left behind.

In places like Canfranc International Railway station in Spain, the grandeur of the architecture still reflects its intended use as an international hub connecting Spain and France. Today, its deserted platforms are silent storytellers of a fallen age and an emblem of the fleeting nature of human creations.

From the deserted platforms in Detroit's Michigan Central Station, once a bustling node of industrial might, now a desolate cathedral of progress past, to India's Dhanushkodi Railway Station, swallowed by a cyclone and reclaimed by nature, these empty stations echo the sentiments that progress always has a price, often paid by the footprints we leave behind.

12.3. The Dialogue with Modern Transport Infrastructure

Our modern transport infrastructure represents the apex of human innovation. Superfast trains, expansive metro systems, and interlinked transit networks define today's era. The energy in these thriving hubs is palpable, the very antithesis of the abandoned railways they supplanted.

Still, as we journey through the sophisticated labyrinth of modern rail transit, we cannot overlook the importance of these 'ghosts of the past.' The abandoned stations around us are crucial links in understanding our journey – a journey that started with the iron horse's crude thundering rhythm and has led us to the magnetic levitation trains' silent soar.

12.4. Lessons from Our Ghostly Past

The lessons these railway stations let us unearth are not only crucial for understanding our past but they also offer valuable insights for our future.

In their ruined facades and crumbling platforms, we learn a poignant lesson about obsolescence. Will our innovative creations today stand the test of time, or will they also fall into disrepair and be deserted tomorrow?

We learn about the relentless march of progress, how it brings both creation and destruction. The once bustling stations turned silent, positions us to ponder about the transient nature of relevance and utility.

Another teachable moment from these ghostly platforms is their co-existence with nature. Stations like New York's High Line and Paris's Petite Ceinture were revived, not as functional hubs, but as rail-trails where nature has been invited to reclaim these abandoned places.

12.5. Conclusion: Echoes of Our Journey

In conclusion, as we step into the era of hyperloops, supersonic travels, and AI-enabled transport, we must remember to periodically turn back and revisit these ghosts of the past, residing in our forgotten railway stations. By doing so, we may yield valuable lessons about the cycles of progress, obsolesence, and reinvigoration.

These abandoned railway stations, chapters in our collective architectural heritage, are places for reflection, for understanding the past, and for framing our ambitions for the future. They will continue to stand as testament to our journey, whispering lessons to those who choose to listen. Therein lies their romance, their allure, their unshakable relevance.